HOW NASA DISCOVERED EARTH 2.0 AND WHY WE'RE ABOUT TO FIND MORE

Finding Alien Sphere

Exploring the Universe's Hidden Worlds and Humanity's Quest for Life Beyond Our Planet

Scott W. Diego

Copyright © Scott W. Diego, 2024.

All rights reserved. No part of this publication may be reproduced, distributed, or transmitted in any form or by any means, including photocopying, recording, or other electronic or mechanical methods, without the prior written permission of the publisher, except in the case of brief quotations embodied in critical reviews and certain other noncommercial uses permitted by copyright law.

Table of Contents

Introduction 3
Chapter 1: The Dawn of Exoplanet Discovery 6
Chapter 2: Exploring Strange and Familiar Worlds 14
Chapter 3: The James Webb Space Telescope and New Horizons 24
Chapter 4: Life as We Don't Know It – Possibilities Beyond Earth 35
Chapter 5: Challenges in Detecting and Exploring Exoplanets 42
Chapter 6: Philosophical and Societal Implications of Discovering Earth 2.0 49
Conclusion 57

Introduction

For centuries, humanity has looked up at the night sky, eyes fixed on the stars, wondering if someone, somewhere out there, might be looking back. This curiosity has fueled myths, inspired art, and shaped our collective imagination, but it has also become the driving force behind one of the most ambitious quests of modern science: the search for life beyond Earth. The universe, vast and enigmatic, holds more mysteries than answers, and in the face of its boundless expanse, we are left to ponder if other worlds, perhaps even other versions of our own Earth, might be hidden among the stars.

This exploration isn't just a philosophical pursuit; it has profound scientific and societal implications. Finding another planet like Earth—one that hosts water, a stable atmosphere, and possibly life—could redefine our understanding of biology, evolution, and perhaps even consciousness itself. Such a discovery would remind us that Earth, with all its diversity of life, might not be a singular exception but part of a larger, shared cosmos. And if we're not alone, what would it mean for our identity as a species? How would it reshape our scientific theories, philosophies, and perhaps even the way we treat each other on this shared, fragile planet?

In the pursuit of these questions, NASA has emerged as a beacon of human curiosity and innovation. With groundbreaking missions and telescopes, it has made incredible strides in exploring beyond our solar system, driven by the hope of identifying what many call Earth 2.0. NASA's journey has spanned decades, from initial forays into nearby planets to the latest advancements in exoplanet detection, using sophisticated tools that can detect the faintest signs of distant worlds. Through its dedication, NASA has enabled humanity to peer deeper into space than ever before.

Central to this exploration is the concept of the "Goldilocks Zone," a term scientists use to describe the range around a star where conditions are just right for liquid water to exist. This zone isn't too hot, nor too cold, making it the prime candidate for finding planets with Earth-like conditions. By studying planets within this range, researchers aim to identify not only habitable worlds but also clues about how life might emerge elsewhere. NASA's commitment to this search highlights the value of exoplanetary study—not merely as a search for a backup to Earth but as a way to unlock the mysteries of our existence and to extend our reach into the unknown.

As we continue our journey through this book, we'll follow NASA's footsteps and look at how it has pushed the boundaries of what's possible, leading us ever closer to discovering if we are truly alone or part of a vast cosmic neighborhood. This journey is about finding answers, but it's also about understanding the questions that drive us to explore in the first place.

Chapter 1: The Dawn of Exoplanet Discovery

The idea of worlds beyond our solar system has intrigued astronomers for centuries. Early thinkers speculated that, given the abundance of stars in the night sky, some of them must host their own planets. Yet, without advanced instruments to confirm it, these ideas remained largely speculative. Early astronomers studied the stars closely, noting their behavior, brightness, and movement, but the existence of planets beyond our solar system—what we now call exoplanets—remained unproven. As the field of astronomy advanced, scientists developed more sophisticated theories about planetary formation and the possibility of Earth-like worlds elsewhere, setting the stage for what would eventually become a groundbreaking era in space exploration.

The first breakthrough came in the late 20th century, marking a pivotal shift from theory to evidence. In 1992, astronomers made the first confirmed detection of exoplanets orbiting a pulsar, a highly magnetized neutron star. This discovery was unexpected, as pulsars are remnants of supernova explosions, and the intense radiation they emit was thought to make it unlikely for planets to exist in their vicinity. However, the evidence was undeniable: slight variations in the pulsar's signals indicated the gravitational pull of orbiting objects, which turned out to be planets. This discovery shattered previous assumptions, proving not only that planets existed beyond our solar system but that they could form and survive in environments once deemed inhospitable.

The next major leap came in 1995, when astronomers identified 51 Pegasi b, a gas giant orbiting a Sun-like star located about 50 light-years away. This detection was monumental as it was the first time a planet was found orbiting a star similar to our own, confirming that our solar system was not unique. It opened the door to a new era of exoplanet research, as scientists realized that with the right tools, we could identify distant planets around other stars. Following this discovery, exoplanet research gained momentum, and scientists began developing new methods and technologies to detect these distant worlds.

In 2009, NASA launched the Kepler Space Telescope, a mission that would redefine our understanding of the cosmos and forever change the field of exoplanet research. Kepler was designed specifically to locate planets beyond our solar system by observing a fixed patch of stars in our galaxy. Equipped with a powerful camera, Kepler's mission was to detect the minute dimming of a star's light as a planet passed in front of it—a method known as the transit method. By observing these dips in brightness, Kepler could estimate the size of the planet, its distance from the star, and its orbital period. This approach allowed scientists to gather detailed information about exoplanets, including some in the so-called "habitable zone" where conditions might support life.

Over the course of its mission, Kepler collected data on hundreds of thousands of stars, uncovering thousands of exoplanets and fundamentally changing our perspective on the prevalence of planets in the galaxy. Among Kepler's most significant contributions was the realization that planets are common, with most stars likely hosting at least one. This shift in understanding suggested that our galaxy alone could contain billions of planets, including many that might have conditions similar to Earth's.

Kepler's discoveries extended beyond quantity; they revealed the incredible diversity of planetary systems. Some systems contained planets tightly clustered around their stars, while others featured super-Earths—planets larger than Earth but smaller than gas giants. Kepler also found planets in binary star systems and other configurations once thought unlikely to support planetary formation. This variety not only expanded our understanding of planetary formation but also hinted at the potential for diverse life-supporting environments beyond our solar system.

Though Kepler's mission ended in 2018, its legacy endures. The data it collected continues to be analyzed, revealing new exoplanets and offering insights into the structure of the galaxy. Kepler taught us that planets are not just occasional outliers but are abundant throughout the cosmos. Its findings provided a foundation for future missions and inspired the scientific community to dream even bigger. In its wake, the search for Earth 2.0 no longer seemed like science fiction but rather a scientific pursuit within our reach, setting the stage for the next generation of telescopes and space missions.

In the search for planets that might support life, scientists focus on a specific region around a star known as the "Goldilocks Zone," or habitable zone. Named for the fairy tale in which Goldilocks seeks a porridge that is "just right," this zone represents a delicate balance. In this region, the conditions are not too hot and not too cold but just right for liquid water to exist on a planet's surface. This potential for liquid water is central to the search for life, as water is a critical ingredient for life as we know it. Within this zone, planets might maintain stable atmospheres and temperatures that allow for the chemical processes necessary for life.

The importance of the Goldilocks Zone lies in its relationship to temperature and atmospheric pressure, both of which are key to sustaining liquid water. If a planet orbits too close to its star, temperatures soar, causing any water to evaporate or remain in vapor form, while planets too far from their star are likely to have frozen, icy surfaces. But in the Goldilocks Zone, there exists the right balance where temperatures remain moderate, allowing water to stay liquid and potentially creating a stable environment where life might flourish.

Earth's location in our solar system provides an ideal example of the Goldilocks Zone at work.

Positioned at an optimal distance from the Sun, Earth avoids the extreme temperatures found on planets like Venus, which is too close and searingly hot, and Mars, which is further out and much colder. Earth's placement allows for a climate range that supports oceans, rivers, and lakes—environments that have been essential for the evolution and sustenance of diverse life forms. This positioning within the habitable zone means that Earth has experienced conditions stable enough over billions of years for complex ecosystems to develop and thrive.

However, the Goldilocks Zone is not a fixed distance; it can vary depending on the type and size of the star. Smaller, cooler stars have closer habitable zones, while larger, hotter stars push their habitable zones further out. This variability means that the concept of habitability isn't limited to Earth-like solar systems but can extend to a wide variety of star systems, each with its unique Goldilocks Zone. This adaptability allows scientists to explore a broader range of potential homes for life beyond our solar system, knowing that while conditions may differ, the essentials for life could still be present.

Identifying planets within these habitable zones offers the best chance of finding life as we understand it. The discovery of such planets, thanks to missions like Kepler, has fueled optimism that Earth-like worlds may not be as rare as once thought. With each new planet located within its star's Goldilocks Zone, we take a step closer to understanding if and where life might exist elsewhere in the universe. The Goldilocks Zone thus remains a guiding principle in our exploration of the cosmos, helping us pinpoint where we might one day find a second Earth, teeming with life and waiting to be discovered.

Chapter 2: Exploring Strange and Familiar Worlds

The variety of planets beyond our solar system has revealed a cosmos rich in complexity and diversity. As scientists observe exoplanets, they encounter a range of planetary types, each with distinct characteristics that hint at the uniqueness of different star systems. Some exoplanets resemble the gas giants of our solar system, such as Jupiter and Saturn, with thick atmospheres of hydrogen and helium swirling around a dense core. These gas giants are often massive, with sizes that dwarf Earth, and they tend to orbit further from their stars, where their gaseous envelopes can remain stable without being stripped away by solar radiation.

Then, there are rocky planets, which bear a closer resemblance to Earth, Mars, Venus, and Mercury. These planets are generally smaller, with solid surfaces composed of rock and metal. Rocky exoplanets, especially those within the Goldilocks

Zone, have captured significant scientific interest, as their composition may allow for atmospheres, liquid water, and potentially life-supporting environments. While some of these rocky planets are "super-Earths" (larger and more massive than Earth but still within a similar structure), others might be smaller and similar in scale to Mars.

Ice worlds add another intriguing dimension to the exoplanet catalog. Found farther from their stars, these planets often exist in extremely cold regions, where water and other compounds freeze into solid forms. These icy exoplanets sometimes have thick layers of frozen water or other substances like methane or ammonia surrounding a solid core. Although seemingly inhospitable, ice worlds raise interesting questions about what kind of life might survive under such conditions, perhaps even beneath thick layers of ice as seen on some moons within our own solar system, like Europa and Enceladus.

As astronomers continue to identify these diverse planets, they have also discovered unique planetary systems that challenge our previous notions about how solar systems are organized. One such system is TRAPPIST-1, a compact collection of seven rocky planets orbiting a cool red dwarf star located about 40 light-years away. Remarkably, all seven planets in the TRAPPIST-1 system are similar in size to Earth and Venus, and three of them are located within the star's habitable zone. What makes TRAPPIST-1 especially fascinating is the proximity of these planets to each other; their orbits are so close that an observer on one planet could potentially see neighboring planets in the sky with the naked eye.

This "mini solar system" structure, with multiple planets packed tightly around their star, offers tantalizing possibilities. If any of the TRAPPIST-1 planets harbor life, interplanetary travel might be feasible within this system, allowing life forms on one planet to visit others. For scientists,

TRAPPIST-1 is a reminder that other solar systems can be very different from our own and that habitable conditions might not be as rare as once thought. Observations of this system hint that densely packed orbits around small, cool stars could be common throughout the galaxy, increasing the odds of finding other compact systems with multiple Earth-like worlds.

This diversity in exoplanets and their systems expands our understanding of what might be possible across the cosmos. While our solar system offers one example of planetary arrangement, discoveries like TRAPPIST-1 suggest that the universe holds countless other configurations, each potentially harboring the conditions for life. With each new discovery, we learn that our planet and solar system are just one of many variations in the rich tapestry of the galaxy, each with its potential for supporting life and offering new mysteries to explore.

Our solar system stands out as a distinctive arrangement in the cosmos, marked by a spacious layout and a variety of planet types that span a wide range of distances from the Sun. Beginning with the rocky, terrestrial planets—Mercury, Venus, Earth, and Mars—nearest to the Sun, and extending outward to the gas giants, Jupiter and Saturn, and the icy giants, Uranus and Neptune, our solar system offers a carefully spaced progression of planetary types. This arrangement allows each planet to orbit comfortably without the gravitational chaos that might arise from closer proximity. In addition, our solar system's planets exhibit remarkable diversity, with solid, rocky worlds on the inside and gaseous and icy giants on the outer edges, creating a balanced structure where planets are spaced far enough apart to avoid major orbital disturbances.

In contrast, many exoplanetary systems discovered thus far look quite different. In these systems, planets often orbit much closer to their stars than we see in our own solar system, leading to densely packed orbits where planets have little "personal space" and sometimes even influence each other's orbits through gravitational interactions. These compact systems, especially around red dwarf stars, frequently feature planets all of similar sizes, often rocky or slightly larger "super-Earths." This pattern of tightly clustered planets differs sharply from the solar system's arrangement, where planets are widely spread out and vary significantly in both composition and size.

One example of such a compact system is TRAPPIST-1, where seven Earth-sized planets orbit within a distance that could fit comfortably inside Mercury's orbit around the Sun. In systems like these, planets are close enough that they may gravitationally interact in ways that can alter their orbits and even affect their atmospheres over time. Additionally, planets in these systems are often tidally locked, meaning one side perpetually faces the star while the other remains in darkness, a stark contrast to Earth's rotation, which enables day-night cycles and helps stabilize climate.

The solar system's arrangement is thus relatively rare compared to the compact, clustered patterns seen elsewhere, and this uniqueness may have contributed to Earth's favorable conditions for life. With sufficient distance from the Sun to avoid extreme heat and stable orbits to maintain a consistent climate, Earth has thrived in a balanced, spacious system. While compact systems offer intriguing possibilities, especially for interplanetary travel, our solar system's design provides a model of stability and diversity that might be just as special in the galaxy.

Among the many exoplanets discovered, a few stand out as intriguing Earth-like candidates that could, in theory, support life or at least maintain similar conditions to those on Earth. These planets have been identified within the habitable zones of their respective stars—where temperatures might allow liquid water to exist—and offer a glimpse into the possibility of finding a true "Earth 2.0."

One of the most famous candidates is **Kepler-452b**, a planet that has drawn significant interest due to its similarities with Earth. Kepler-452b is a "super-Earth," about 60% larger than our planet, and it orbits a star very similar to our Sun. This planet resides in its star's habitable zone, where temperatures are likely warm enough to support liquid water. Scientists believe Kepler-452b has a rocky composition, similar to Earth, and its position within the habitable zone has led to speculation that it might have the potential to host an atmosphere and possibly even support basic life forms. This discovery was monumental because it suggested that Earth-like planets could exist around Sun-like stars, raising the tantalizing prospect of worlds with conditions strikingly similar to our own.

Another intriguing exoplanet is **Kepler-186f**, which is often hailed as one of the most Earth-like planets discovered to date. Kepler-186f is just about 10% larger than Earth and orbits within the habitable zone of a smaller, cooler red dwarf star. What makes Kepler-186f particularly compelling is its potential for seasonal changes and a stable climate, much like Earth. Scientists theorize that this planet has a similar axial tilt, meaning it could experience seasons and a consistent climate. Additionally, its position around a red dwarf star suggests that it may have had time to develop stable environmental conditions, as these stars can burn for billions of years. Although much about Kepler-186f remains speculative, its discovery has fueled hope that Earth-sized, potentially habitable planets may be more common than previously thought.

Beyond these individual finds, the Kepler mission has revealed a broad array of **"super-Earths" and Goldilocks planets** that add to the list of intriguing candidates. Super-Earths are planets larger than Earth but smaller than Neptune, often with rocky surfaces or thick atmospheres, depending on their proximity to their stars. Many of these super-Earths are found in their stars' habitable zones, with conditions that may allow for liquid water and, potentially, life. For instance, planets like Kepler-62f and Kepler-22b are thought to have stable orbits within habitable zones, with potential for atmospheres and oceans.

These discoveries offer promising leads in the search for habitable worlds. Each exoplanet observed in the Goldilocks Zone brings us closer to understanding whether life-supporting planets are unique to Earth or a common feature in the galaxy. As researchers continue to explore these planets, we gain new insights into the conditions that might allow life to flourish elsewhere, strengthening the hope that an Earth 2.0 could be waiting to be discovered among the stars.

Chapter 3: The James Webb Space Telescope and New Horizons

The James Webb Space Telescope (JWST) represents a leap forward in our ability to observe the universe, building on decades of technological development and vision. Conceived as a successor to the Hubble Space Telescope, JWST is designed to probe deeper into space, uncovering secrets of the cosmos that have remained hidden from view. Its mission is ambitious: to capture light from some of the first stars and galaxies formed after the Big Bang, study the formation of stars and planetary systems, and investigate the atmospheres of exoplanets for possible signs of life. Each of these objectives requires cutting-edge technology and precision that JWST delivers through its innovative design.

A defining feature of JWST is its massive segmented mirror, spanning 6.5 meters in diameter—over twice the size of Hubble's mirror. This gold-coated mirror is composed of 18 hexagonal segments that can adjust to achieve perfect focus. The large size of the mirror allows JWST to collect far more light than previous telescopes, giving it unparalleled sensitivity to faint and distant objects. Unlike Hubble, which primarily operates in the visible spectrum, JWST is designed to observe in the infrared range. This capability enables it to detect objects obscured by clouds of gas and dust, such as newly forming stars and planets, and to observe distant galaxies whose light has redshifted into the infrared as it travels across vast cosmic distances.

Another significant aspect of JWST is its location. Unlike Hubble, which orbits relatively close to Earth, JWST operates from a position 1.5 million kilometers away, at the second Lagrange point (L2). This distance helps JWST maintain a stable orbit that keeps it aligned with Earth as the planet orbits the Sun, providing an uninterrupted view of space. Additionally, its placement reduces interference from Earth's heat and light, which is crucial for observing in the infrared. JWST's position at L2 also allows for the deployment of a large, five-layered sunshield that spans nearly the size of a tennis court, keeping the telescope's instruments cool by blocking heat from the Sun, Earth, and Moon. This cooling is essential, as infrared observations require instruments to operate at extremely low temperatures to detect the faintest heat signals from distant celestial objects.

With its advanced capabilities, JWST brings a unique vision to the exploration of the cosmos. Its design, combining a large mirror, infrared vision, and a distant orbit, allows it to see further back in time than any previous telescope, capturing images and data that will help unravel the history and structure of the universe. This combination of technologies is precisely what makes JWST a powerful tool in the search for Earth-like planets and signs of life, as it can analyze the atmospheres of exoplanets and detect chemical signatures that might indicate biological processes. JWST's potential is unmatched, promising to deepen our understanding of everything from star formation to the potential for life on distant worlds. As it opens its eyes to the universe, JWST offers humanity an unprecedented view of our cosmic origins and the vast possibilities that lie beyond our solar system.

The James Webb Space Telescope (JWST) has transformed our approach to studying exoplanets, enabling scientists to probe regions of the cosmos that were previously hidden or too distant to observe with clarity. One of JWST's most groundbreaking features is its use of infrared technology, which allows it to "see" through dense cosmic dust clouds that would otherwise obscure distant stars, galaxies, and planets. Unlike visible light, which can be easily blocked by dust and gas, infrared light can penetrate these barriers, providing clear images of regions where stars and planets are actively forming. This capability lets JWST examine protoplanetary disks—clouds of gas and dust around young stars where planets take shape—and gather insights into the early stages of planetary evolution.

JWST's infrared sensors are also key to its ability to observe the universe's earliest moments, a task often described as "seeing backward in time." Because light from distant galaxies has traveled billions of years to reach us, JWST captures these celestial objects as they appeared in the past. As the universe expands, light from ancient galaxies stretches into longer, redder wavelengths, shifting from visible into the infrared spectrum. JWST's ability to capture this redshifted light means it can observe some of the first stars and galaxies that formed shortly after the Big Bang, offering a window into the origins of structure in the universe. Through these observations, scientists hope to uncover how the earliest galaxies formed and evolved, leading to the creation of stars, planets, and ultimately the conditions that support life.

In addition to these capabilities, JWST is pioneering advancements in atmospheric analysis for exoplanets, taking exoplanet research into unprecedented territory. When a planet passes in front of its star—an event known as a transit—the star's light filters through the planet's atmosphere, if one exists, leaving behind spectral fingerprints. JWST can detect these subtle changes in light and analyze the composition of the planet's atmosphere with remarkable precision. By identifying the specific wavelengths absorbed by elements and molecules, scientists can determine whether the atmosphere contains water vapor, carbon dioxide, methane, or even potential biosignatures—chemical markers that could indicate biological processes.

This level of atmospheric analysis goes far beyond the capabilities of previous telescopes, bringing scientists closer to determining the habitability of distant worlds. For the first time, JWST can search for signs of complex molecules or gases that might signal the presence of life, such as oxygen or methane in the right proportions. It can also distinguish between different types of atmospheres, offering clues about the planet's surface conditions and climate, and helping researchers evaluate whether it might have oceans, clouds, or even seasonal changes.

By combining these advancements, JWST is reshaping exoplanet research and expanding our understanding of where life might exist beyond Earth. The telescope's ability to peer through dust, observe ancient galaxies, and analyze exoplanet atmospheres sets a new standard in the search for habitable worlds. It not only enhances our knowledge of exoplanets but also helps answer fundamental questions about how planets and life come to be. Through JWST's eyes, we are gaining unprecedented insights into the building blocks of life across the universe, and we are closer than ever to discovering whether we are alone in the cosmos or part of a vast interstellar community.

The James Webb Space Telescope (JWST) represents a leap forward in the search for a second Earth, a world that could potentially mirror the conditions of our own. With its advanced observational power, JWST can identify and examine exoplanets with a precision that was unimaginable just a few years ago. By targeting planets in the habitable zones of distant stars—regions where conditions may allow for liquid water and, by extension, life—JWST brings us closer to finding a true Earth 2.0.

One of JWST's most significant contributions to this search lies in its ability to analyze exoplanet atmospheres for markers that could suggest habitability. The telescope's infrared sensors are finely tuned to detect a range of gases, including water vapor, carbon dioxide, methane, and other compounds that might hint at biological processes. By capturing the light that filters through a planet's atmosphere during a transit, JWST can produce a detailed "chemical fingerprint" of the gases present. This analysis offers scientists insights into a planet's surface conditions, climate stability, and the potential for life-supporting environments. In doing so, JWST can distinguish between barren worlds and those that might host life, honing in on the most promising candidates.

JWST's capabilities are particularly useful for following up on promising exoplanets first detected by the Kepler Space Telescope. Kepler identified thousands of potential planets, including Earth-like candidates such as Kepler-452b and Kepler-186f, which orbit within their stars' habitable zones. However, Kepler's primary role was to locate planets by observing slight dimming as they passed in front of their stars; it was not equipped to analyze atmospheric details. Now, JWST can take these findings a step further, using its powerful sensors to study these planets in depth, examining their atmospheres for biosignatures or other signs that life could exist there. In many ways, JWST acts as a bridge between discovery and understanding, turning Kepler's list of exoplanets into a detailed map of worlds that might sustain life.

The potential discoveries JWST could make are as exciting as they are profound. With its ability to detect atmospheric compositions, JWST can, in theory, find evidence of chemical imbalances that suggest biological activity—such as an unusual mix of oxygen and methane that would be difficult to explain by non-biological means. If JWST detects such patterns, it would mark one of the most groundbreaking discoveries in human history, as we would have evidence, however indirect, that life might exist elsewhere in the galaxy.

The search for Earth 2.0 is no longer just a theoretical endeavor; JWST has brought us to the threshold of concrete possibilities. Each planet it studies, each atmosphere it samples, and each spectral signature it analyzes brings us closer to a world that might reflect our own. With JWST's unprecedented observational power, scientists are optimistic that the discovery of an Earth-like planet, complete with the conditions for life, may be within reach sooner than we think. The journey to find Earth 2.0 is unfolding in real time, and JWST stands at the forefront, revealing the tantalizing potential that another habitable world may be out there, waiting to be discovered.

Chapter 4: Life as We Don't Know It – Possibilities Beyond Earth

Our search for extraterrestrial life is largely guided by the conditions that sustain life on Earth—water, moderate temperatures, and a mix of gases like oxygen and nitrogen. But by focusing on these familiar requirements, we may be narrowing our view, potentially missing forms of life that don't fit our Earth-based definitions. Expanding our perspective means considering life that could survive in environments radically different from those we know, where elements like methane or hydrogen could take the place of water, and life could thrive under pressures or temperatures that would be uninhabitable for us.

One example of such alternative environments might be a planet with methane seas. On Saturn's moon Titan, for instance, rivers and lakes of liquid methane and ethane flow across its icy surface, a phenomenon unimaginable on Earth. Although these conditions would be hostile to terrestrial life, it raises the possibility that some forms of life could be adapted to use methane as a biological solvent instead of water. Another potential setting for alien life is within clouds of hydrogen or in high-pressure atmospheres like those found on super-Earths. While these scenarios seem inhospitable by Earth standards, they might offer the stability needed for unique biochemical processes, creating life forms that operate on principles we have yet to imagine.

The idea of life existing in extreme environments isn't as far-fetched as it might seem. On Earth, certain resilient organisms—called extremophiles—thrive in places once thought uninhabitable. Among the most famous of these extremophiles are tardigrades, tiny, resilient creatures capable of surviving intense heat, extreme cold, and even the vacuum of space. Tardigrades can endure dehydration for years, recover from radiation exposure, and survive pressures that would crush other organisms. This adaptability has made them a powerful symbol of life's potential to thrive beyond familiar settings.

By studying extremophiles like tardigrades, scientists are exploring what life might look like on alien worlds where conditions are harsher than any found on Earth. The resilience of these organisms suggests that life may be more flexible and resourceful than we currently understand. If tardigrade-like organisms exist elsewhere in the universe, they could be thriving in places that lack Earth-like warmth, sunlight, or even liquid water. Perhaps they exist within layers of frozen ice, under high-pressure atmospheres, or even floating in clouds of gas.

Broadening our view of life's requirements allows us to consider new possibilities and expands the scope of our search for extraterrestrial life. Instead of limiting ourselves to planets in the traditional "Goldilocks Zone," we might look at icy moons, methane-rich worlds, or planets enveloped in clouds of gas, where life could adapt in ways we have yet to comprehend. In a universe so vast and varied, our understanding of biology may only scratch the surface of what is possible, and it is within these extreme, unconventional environments that we might one day discover forms of life that redefine our very notion of existence.

For centuries, human imagination has populated the cosmos with diverse forms of alien life. Stories of extraterrestrials have ranged from humanoid beings who look eerily like us to surreal and monstrous creatures far beyond the bounds of earthly biology. These visions, crafted through myth, literature, and cinema, reflect our own desires, fears, and curiosity about the unknown. While popular culture often depicts aliens as either familiar or frightening, scientific speculation offers a much broader and, perhaps, stranger view of what alien life could look like.

In fiction, aliens are often portrayed as human-like—bipedal, intelligent beings with recognizable faces and expressions. This "humanoid" model is comforting because it assumes a common evolutionary path and cognitive similarity. However, scientists believe this view is likely limited by our tendency to project human traits onto the unknown. Evolution on other planets could follow entirely different trajectories, producing life forms that bear little resemblance to anything on Earth. The diversity of species on our own planet hints at the vast possibilities that might emerge on worlds with different climates, atmospheres, and gravitational forces. Alien life may be as varied as Earth's own organisms, but adapted to habitats beyond our comprehension.

Scientific theories about alien life have grown increasingly bold with recent discoveries. Some researchers speculate about creatures that could survive in the thick atmospheres of gas giants, where organisms might float through the clouds, feeding on chemical reactions instead of organic matter. Others envision life forms that thrive in methane-rich lakes, using chemical compounds as a basis for life in place of oxygen. These ideas remind us that extraterrestrial life may not need to resemble anything familiar to us at all; it could be microbial, plant-like, or something entirely different—a complex, self-sustaining system that challenges our definitions of life.

New discoveries in the fields of exoplanet research and astrobiology have only deepened these possibilities. As scientists uncover more about extremophiles on Earth—organisms that live in boiling vents, acidic pools, or the frozen permafrost—they are constantly reminded that life is resilient and adaptable. These revelations expand our concept of what an alien could be and where it might live. With each new finding, from water ice on Mars to organic molecules on distant moons, our expectations evolve, making room for life forms that could be radically different from any we've imagined.

As our understanding of the cosmos advances, the line between science and science fiction begins to blur. The possibility of discovering life in forms beyond our current comprehension encourages us to think creatively and be prepared for the unexpected. Perhaps the universe is filled with life forms as complex and varied as those in our imagination, from civilizations far advanced beyond our own to simple organisms existing in the shadows of distant moons. Each new discovery holds the potential to reshape our understanding of life itself, reminding us that the universe may be more mysterious—and more populated—than we ever dared to dream.

Chapter 5: Challenges in Detecting and Exploring Exoplanets

The search for alien worlds and signs of life beyond Earth faces significant technological challenges, many of which stem from the sheer distances and faintness of these distant planets. One of the greatest hurdles in exoplanet research is the difficulty of direct imaging. Stars, being vastly larger and brighter than their orbiting planets, emit intense light that often obscures these smaller celestial bodies. Attempting to directly photograph a distant planet from Earth or even from a nearby space telescope is like trying to spot a tiny speck next to a brilliant spotlight. Even advanced telescopes like the James Webb Space Telescope (JWST) struggle to capture clear, isolated images of exoplanets due to this overwhelming glare.

To work around the limitations of direct imaging, scientists have turned to indirect methods, one of the most powerful being atmospheric composition analysis. Instead of attempting to photograph these distant worlds, researchers analyze the light from the planets' host stars, especially during transits, when a planet crosses in front of its star as seen from our perspective. During this brief period, some of the starlight passes through the planet's atmosphere (if it has one), carrying with it subtle changes that reflect the atmosphere's chemical makeup. By examining the specific wavelengths of light absorbed by the atmosphere, scientists can identify the presence of various gases, such as water vapor, carbon dioxide, methane, and even oxygen.

This technique, called transmission spectroscopy, offers a practical means of "seeing" what might be in a planet's atmosphere without requiring a direct image. Although it cannot reveal visual details of an exoplanet, it provides an invaluable set of clues about its composition and potential for life. By detecting certain combinations of gases—particularly those that are reactive and would typically require replenishment, like oxygen and methane—scientists can search for "biosignatures," or signs that biological processes might be at work. For instance, on Earth, methane is quickly broken down by sunlight, so its presence in our atmosphere is continually refreshed by biological activity. Discovering a similar pattern on another planet could hint at the existence of life.

Atmospheric composition analysis is an innovative solution to the limitations posed by the vastness of space and the brightness of stars, giving scientists a way to study distant planets indirectly. Although direct imaging may remain an elusive goal for now, tools like the JWST and other future telescopes are gradually refining our view, pushing the boundaries of what's possible in exoplanet research. As technology continues to advance, we inch closer to identifying not just the physical characteristics of these alien worlds but also the potential signs of life they may harbor, all without needing to capture a single photograph.

In the search for exoplanets and potential signs of life, the role of data is crucial. Modern telescopes like the James Webb Space Telescope (JWST) and previously the Kepler Space Telescope collect vast amounts of information, observing thousands of stars and planets across vast regions of space. The volume of data these instruments generate is staggering, and processing it requires sophisticated methods. Every tiny dip in a star's brightness or subtle shift in light spectrum could indicate a potential exoplanet, a change in atmospheric composition, or even hints of biological activity. But sifting through these massive data sets to identify meaningful patterns is a task far beyond human capacity, leading scientists to harness the power of artificial intelligence (AI) and machine learning.

AI algorithms, trained to recognize specific features and anomalies, can analyze astronomical data at remarkable speeds and with increasing accuracy. Machine learning models can be taught to detect the unique signatures of planets transiting in front of their stars, classify the types of planets observed, and even identify potential atmospheric biosignatures. These algorithms are not only faster than traditional data processing but can also learn from previous discoveries, improving their ability to spot new candidates with each analysis. Machine learning is particularly valuable for studying spectral data, where it can highlight the presence of gases like methane or water vapor in exoplanet atmospheres. By leveraging these tools, scientists can accelerate the discovery process, identifying planets and atmospheric conditions that warrant closer investigation.

Yet, even as we refine our techniques for identifying exoplanets, the challenge of reaching these distant worlds remains a fundamental limitation. Currently, interstellar travel—physically visiting planets beyond our solar system—is beyond the reach of existing technology. The vast distances involved are staggering; even the closest exoplanets, like those in the Proxima Centauri system, are over four light-years away. With current propulsion technology, it would take tens of thousands of years to reach even the nearest stars.

However, scientists and engineers are exploring theoretical advancements that could one day make interstellar travel feasible. Concepts such as light sails, which use laser-driven propulsion to push small, lightweight spacecraft at a fraction of light speed, offer one potential solution. The Breakthrough Starshot initiative, for instance, aims to launch tiny probes equipped with light sails to the Alpha Centauri system, potentially reaching these neighboring stars within a few decades. Other ideas, like nuclear fusion propulsion or antimatter drives, remain speculative but promise the possibility of faster travel through space if developed. Even more futuristic concepts, such as warp drives, which would theoretically allow a spacecraft to "warp" space-time and achieve faster-than-light travel, are being studied at a theoretical level, though they remain far from practical implementation.

While physical exploration of exoplanets may lie in the distant future, the combination of advanced data analysis and theoretical propulsion research represents humanity's best hope for one day bridging the gap between us and distant star systems. As AI and machine learning help us find and understand potentially habitable worlds, the groundwork is being laid for future technologies that might one day allow us to reach them. In the meantime, our telescopes and data-processing tools continue to draw back the curtain on these alien worlds, bringing us closer to answering some of the most profound questions about life and our place in the universe.

Chapter 6: Philosophical and Societal Implications of Discovering Earth 2.0

The question of whether humanity is ready for first contact with extraterrestrial life is as complex as it is profound. For centuries, people have speculated about life beyond Earth, and yet, despite all our advancements, the actual discovery of alien life—especially intelligent life—would likely provoke a range of reactions that extend far beyond scientific excitement. For some, the confirmation of extraterrestrial existence might be awe-inspiring, a moment of wonder that reaffirms the vastness and diversity of the cosmos. But for others, it could lead to feelings of existential unease, prompting questions about our place in the universe and even fears about potential threats or challenges to humanity.

The societal impact of such a discovery would likely ripple through every corner of human life, affecting religious beliefs, philosophical inquiries, and humanity's self-perception. Religions around the world offer diverse explanations for creation, life, and the nature of the universe. First contact might challenge these narratives or, alternatively, offer new dimensions of understanding. Some may interpret the discovery of extraterrestrial life as evidence of a shared cosmic creator, expanding their view of divinity to encompass a broader universe. Others may question traditional teachings, feeling compelled to reconcile their beliefs with new realities. Philosophically, contact with alien life would raise questions about the universality of consciousness, intelligence, and moral values, potentially reshaping ethics and expanding the concept of community beyond Earth.

Beyond religion and philosophy, our views on humanity's role and responsibilities could shift dramatically. If life exists elsewhere, especially life similar to ours, it would suggest that Earth may not be as unique or central to the cosmos as once thought. This new perspective could foster a sense of humility, reminding us that we are part of a larger tapestry of life, rather than isolated in the universe. Such a realization might inspire a renewed focus on unity, cooperation, and preservation, not only within our species but toward our planet itself.

This quest to understand our place in the cosmos is deeply tied to humanity's search for belonging. The desire to connect with other intelligent beings speaks to a fundamental curiosity and perhaps even a need to feel part of something greater. The discovery of an Earth 2.0—a world that might harbor life—could have a profound effect on how we view our own planet. Knowing that there are other places in the universe where life might exist could intensify our appreciation for Earth's uniqueness. If such planets exist, but at vast distances or with uncertain habitability, Earth's status as a haven for life may become even more precious in our eyes. This awareness might inspire stronger environmental efforts, recognizing that, despite the allure of other worlds, Earth remains the only world we can call home.

In this way, the search for extraterrestrial life is as much about self-discovery as it is about exploration. As we uncover potential Earth-like planets and push the boundaries of our knowledge, we are driven not just by a desire to find life but by a need to understand our own existence in a broader context. First contact, whether with microbial organisms or advanced civilizations, would be a transformative event, urging us to reflect on our responsibilities to each other, to our planet, and to any life that might share the universe with us. It's a journey that, while scientific at its core, is also deeply personal, reshaping our understanding of what it means to be human in a cosmos filled with possibility.

The question of whether humanity should reach out to alien civilizations is a topic steeped in ethical complexity and caution. While the discovery of extraterrestrial life would be monumental, initiating contact is a different step entirely, one that comes with inherent risks and profound ethical considerations. On one hand, reaching out could offer unprecedented opportunities for knowledge exchange, providing insights into alien technologies, social structures, and perhaps even solutions to challenges we face on Earth. Yet, there are equally compelling reasons to hesitate, as we cannot predict how an advanced civilization might interpret or respond to our communication.

One primary concern is the risk of inadvertently exposing Earth to potential dangers. Much like early explorers who encountered indigenous societies, the arrival of humans might disrupt or harm any alien civilization we come across, and vice versa. History has shown us countless examples of first contact between human cultures that led to devastating consequences, often for the less technologically advanced society. In our eagerness to connect, we may overlook the potential to disrupt alien ecosystems or societal structures in ways we can't foresee. Conversely, an alien civilization might view Earth as a resource to exploit or even as a threat. The implications of contact with a more advanced species are largely unknown, and once made, such a connection could not be undone.

Moreover, there is the ethical question of consent. Any decision to reach out to extraterrestrial beings would inherently be made by a small group of scientists or policymakers, but it would affect all of humanity. Should a decision of such magnitude be undertaken without widespread debate and consideration? The ramifications of contact could extend far beyond the scientific community, impacting cultures, religions, and political systems worldwide. Some argue that any decision to initiate contact should involve a global consensus, recognizing the profound impact such a step would have on everyone on Earth.

Others argue that the curiosity and potential benefits of reaching out outweigh the risks, believing that discovery and connection are fundamental to the human experience. If there are other intelligent beings in the universe, learning from them could accelerate our understanding of science, culture, and the universe itself. Some proponents of contact suggest that humanity has already made its presence known through radio signals and other emissions that have been traveling outward for decades, so attempting to communicate intentionally may be an inevitable next step.

Yet the lessons from history urge caution. For both humanity and any alien life we might encounter, first contact could be transformative but also fraught with misunderstanding and unintended consequences. Like explorers on Earth who encountered vastly different societies, we may find that initial contact brings unforeseen cultural and ethical challenges, potentially altering the trajectory of both civilizations. Humanity stands at a crossroads where the decision to reach out must be weighed carefully, balancing our desire to explore with a respect for the unknown. Ultimately, the choice to communicate is not merely a scientific endeavor but a moral one, demanding reflection on our responsibilities as explorers in a universe that may hold both friends and dangers alike.

Conclusion

As we stand at the frontier of exoplanet exploration, the strides made over the past few decades reveal just how far we have come—and how much lies ahead. Through a combination of groundbreaking technology, visionary projects, and an unwavering curiosity, we have uncovered a growing list of exoplanets, each offering a glimpse into the diversity and complexity of worlds beyond our own. Thanks to NASA's pioneering missions, particularly the Kepler and James Webb Space Telescopes, we have expanded our understanding of the universe, discovering thousands of exoplanets, some of which tantalize us with the possibility of habitability. Kepler set the stage by identifying these distant planets through methods that allowed us to measure their sizes, distances, and compositions. It gave us the foundation for recognizing that planets are common in the galaxy and that Earth-like worlds may be far more prevalent than we once imagined.

Today, the James Webb Space Telescope builds on this legacy, allowing us to examine these distant worlds with unprecedented clarity and detail. Its advanced infrared capabilities enable us to look deeper into space and detect faint signatures in planetary atmospheres, pushing us closer to identifying potential biosignatures. With each new planet it observes, the telescope expands our understanding of how worlds form, evolve, and possibly even host life. As data from JWST continues to arrive, the search for an Earth 2.0 feels more tangible, making the question of life beyond Earth an imminent scientific pursuit rather than a distant dream.

Looking to the future, NASA and other space agencies have ambitious plans for the next phase of exoplanet research. Projects like the upcoming Nancy Grace Roman Space Telescope and the European Space Agency's ARIEL mission aim to deepen our exploration, each with unique instruments designed to survey and characterize exoplanets across the galaxy. Proposed missions, like LUVOIR (Large UV/Optical/IR Surveyor) and HabEx (Habitable Exoplanet Observatory), envision telescopes capable of directly imaging Earth-like planets around distant stars, further advancing our ability to study these worlds in detail. These next-generation missions could reveal atmospheres, climates, and even surface conditions on planets light-years away, taking humanity ever closer to identifying habitable worlds and perhaps even signs of life.

In reflecting on this journey, it becomes clear that the quest to understand life beyond Earth is as much about defining our place in the cosmos as it is about discovery. As we continue to peer into the depths of space, we are reminded of the vastness of the universe and the many mysteries that remain. The search for exoplanets—and ultimately for life—is an exploration of possibility, of reaching out beyond our world to learn about the billions of others that share our galaxy. Each discovery, whether of a barren rock or a potential Earth 2.0, brings us a step closer to answering some of humanity's most profound questions: Are we alone? What else is possible in the tapestry of the cosmos?

This pursuit invites each of us to look up and consider our own role in this cosmic story. As explorers and caretakers of Earth, we have both the privilege and the responsibility to deepen our understanding of the universe, to protect our own planet, and to envision what it means to be part of a larger interstellar community. The discoveries we make today lay the groundwork for a future in which humanity may one day reach beyond its own solar system, exploring, connecting, and discovering in ways we can only begin to imagine. As we continue to gaze toward the stars, we stand on the threshold of answers that may redefine our understanding of life itself, and we are reminded that the universe is a place of infinite possibility, waiting to be explored.

www.ingramcontent.com/pod-product-compliance
Lightning Source LLC
Chambersburg PA
CBHW070130230526
45472CB00004B/1500